YOUR KNOWLEDGE HAS VALUE

- We will publish your bachelor's and
  master's thesis, essays and papers

- Your own eBook and book -
  sold worldwide in all relevant shops

- Earn money with each sale

Upload your text at www.GRIN.com
and publish for free

Yasir Khan

The Reliability and Validity of Real-time PCR Data in Biomedical Sciences

GRIN Publishing

Bibliographic information published by the German National Library:

The German National Library lists this publication in the National Bibliography; detailed bibliographic data are available on the Internet at http://dnb.dnb.de .

Imprint:

Copyright © 2014 GRIN Verlag GmbH
Print and binding: Books on Demand GmbH, Norderstedt Germany
ISBN: 978-3-656-88615-0

This book at GRIN:

http://www.grin.com/en/e-book/288335/the-reliability-and-validity-of-real-time-pcr-data-in-biomedical-sciences

GRIN - Your knowledge has value

Since its foundation in 1998, GRIN has specialized in publishing academic texts by students, college teachers and other academics as e-book and printed book. The website www.grin.com is an ideal platform for presenting term papers, final papers, scientific essays, dissertations and specialist books.

Visit us on the internet:

http://www.grin.com/

http://www.facebook.com/grincom

http://www.twitter.com/grin_com

The reliability and validity of real-time PCR data usage in biomedical sciences

A- Introduction:

i)   Overview of the Project:

Despite a fairly broad implementation and application of real-time PCR, there still exists a vacuum in determining the correct procedures for the examination of quantitative real-time PCR; more explicitly, there is a need to determine appropriate procedures to attain the right kind of statistical treatment. In today's various methods of data analysis, the key statistical inferences are not as exclusive as required like confidence interval. This paper presents and tends to relate four statistical models and approaches on the basis of standard curve method and methods for data analysis.

The first approach developed a multiple regression analysis model for the determination of $\Delta\Delta Ct$ directly from the approximation of interface of gene and treatment paraphernalia. The second approach used the analysis of covariance i.e. ANCOVA model where the derivation of $\Delta\Delta Ct$ could be made through the sequential evaluation and analysis of effects of concurrent variables. The remainder of the models chiefly involves the calculation of ΔCt subsequently connected through the non-parametric comparable Wilcoxon test and a two group T-test. Moreover, a data quality control model was established, which was then applied through the SAS programs determined for all of the aforementioned approaches; analyzed data output was also presented for a sample set.

The SAS programs were used to put forward practical statistical solutions for real-time PCR data while the programs were also utilized to analyze a sample dataset. After a comprehensive analysis conducted through the approaches and models mentioned above, similar results were obtained.

ii)   Guidance from research articles:

While conducting research and studying the retrospect, it was found that the real-time PCR data analysis i.e. the Polymerase Chain Reaction data analysis occupies a pivotal position in the biomedical sciences field. This method can very fairly be regarded as a revolution in the biomedical history. The real-time PCR is not different from the typical PCR, and is solely based on the fundamental principles of the Polymerase Chain Reaction (M. Tevfik Dorak, 2006). This method was primarily initiated roughly a decade ago, and has continued to gain popularity since its very inception.

The real-time PCR is different from the conventional PCR merely in a sense that it emphasizes on the entire process of the chain reaction in spite of the end product. Real-time PCR provides the biomedical scientists with a variety of analytical methods, detection chemistries and a constantly increasing number of platforms. Considering broadly, the real-time PCR is a part of molecular genetics, and is expanding throughout the globe.

The quantitative PCR i.e. qPCR entails a whole range of microbiological fronts comprising of environmental, clinical, food and industrial microbiology (Martin Fillon, 2012). This very visible impact of the qPCR has resulted in making it the center of all the primary attention and research. With a great deal of study, the qPCR technology has evolved a great deal. It has shape shifted from an expensive and complex to an affordable and sensitive technique. Also, with the latest developments and findings, this technique has become faster than ever before. qPCR has evolved into a benchmark for the detection of nucleic acids specifically in microbiology and broadly in biomedical research.

iii)   Purpose of the project:

As an analysis tool, real-time PCR has proved to be one of the fastest and most dependable quantitative methods for the analysis and evaluation of gene expression. Its applications cover a vast paradigm and range broadly from microarray verification and pathogen quantification to drug therapy studies. As a procedure, the PCR has been segmented into three phases listed below;

- Exponential Phase

  This is the primary segmentation of PCR whereby, a significant increase – exponentially – is observed in the product due to the reagents being non-limited. During this phase, the products ideally double in amount provided that the efficiency is a hundred percent.

- Linear Phase

  In this phase, the product increases linearly due to a sudden limitation in the reagents.

- Plateau Phase

  This is the final segmentation of the PCR where the product does not increase due to a depletion of the reagents.

The development of a variety of analysis procedures, methods and programs is based on the fact that the goal of PCR experiments is relative quantification. An in depth investigation of the PCR products can lead to detecting sequence variants (Julie Logan, Kirstin Edwards, Nick A. Saunders, 2009).

iv)   Significance of study:

The real-time PCR technique, as discussed earlier, has become one of the most significantly used and reliable source of genetic analysis. For biomedical sciences in particular, real-time PCR provides the researchers with a fairly simple yet effective way of measuring genetic inferences. The procedure is so artless that it merely requires a piece of template DNA, a thermostabile DNA Polymerase and two primers (M. Wink, Heidelberg, n.d). A great deal of progress has been achieved in experimental sciences through the introduction and evolution of real-time PCR. Within only a short period of time – in the sense of research – real-time PCR has resulted in bringing forward revolutionary findings in molecular biology research and medicinal diagnostics. The real-time PCR is a strong yet unique technology for a quick and comprehensive analysis of the gene expression (Phouthone Keohavong, Stephen G. Grant, 2005). It was only a decade ago that the amplification of specific mRNA became a possibility after the introduction of the PCR technique (Joe O'Connell, 2002).

B- Method:

There exist discrepancies between data quality standards from the two mathematical models for relative quantification of real-time PCR data. The $\Delta\Delta Ct$ method appears to be more rigorous because of the assumption that every reaction must reach an amplification frequency of 2, or in simple language, the amount of product must double during each cycle. Furthermore, the $\Delta\Delta Ct$ method also assumes that for each of the samples under study, the PCR amplification efficiency will be 2. But, such an assumption is strictly contradictory to the effect of various cDNA samples. Now, the efficiency-calibrated method is based on the assumption that different experimental samples share universal amplification frequencies. It is clearly evident that both the mathematical models mentioned here are in contrast to each other.

A data correlation model can be applied to thoroughly examine the data quality. Despite the fact that a correlation analysis between concentration and number of Ct could result in an effective quality control, this can be done in a better way by the analysis of correlation between Ct and the logarithm (base 2) transformed concentration of template, that can result

The reliability and validity of real-time PCR data usage in biomedical sciences

in bringing a substantial simple linear relationship for all the sample concentrations and genes involved.

The input data is grouped as shown in the table below where each classification of gene was named from 1 to 2.

Table 1.0

Replicate	Sample	Gene	Concentration	Ct	Group (Class)
1	Control	Target	10	23.1102	1
2	Control	Target	10	22.9003	1
3	Control	Target	10	22.8972	1
1	Control	Target	2	26.5801	1
2	Control	Target	2	26.2139	1
3	Control	Target	2	26.0606	1
1	Control	Target	0.4	28.1125	1
2	Control	Target	0.4	28.1899	1
3	Control	Target	0.4	27.5949	1
1	Control	Target	0.08	30.2772	1
2	Control	Target	0.08	30.4667	1
3	Control	Target	0.08	30.7571	1
1	Treatment	Target	10	21.7813	2

Figure 1: This figure depicts four classes of four different combinations of gene and sample which include target gene in control sample, reference gene in control sample, reference gene in treatment sample and target gene in treatment sample.

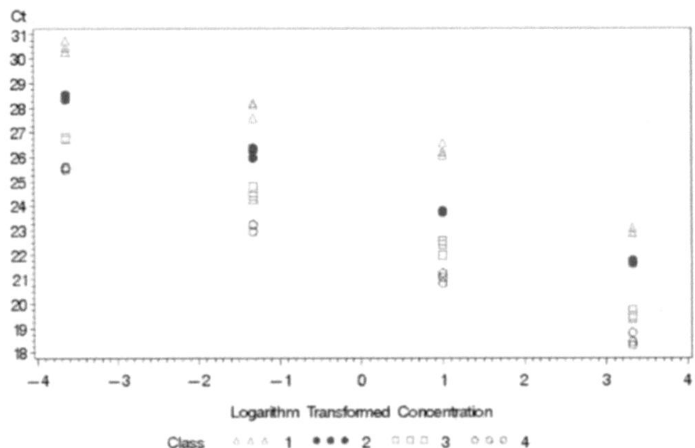

The reliability and validity of real-time PCR data usage in biomedical sciences

Whenever there is a discussion on the production, the terms quality control and quality assurance are of utmost significance. In the case of PCR data analysis, there exists a strong need of data quality control. In the recent times, PCR has been optimized, developed, improved and widely implemented within the known clinical disciplines as elaborated earlier. This is the reason why quality control criteria had to be developed for PCR protocols so as to make it compliant with the international guidelines and GLP i.e. good laboratory practice. The global recognition of PCR assays depends chiefly upon the sensitivity, efficiency, relevance and accuracy of these PCR assays (E. Van Pelt-Verkuil, 2008). Sensitivity, in regard to PCR may be described as the lowest number of target molecules that are specifically detectable by a particular assay. However, there are contradictions as to whether use the term sensitivity or efficiency to best describe this phenomenon.

The PCR – because of its ability to amplify small amounts of nucleic acids – can be used to identify organisms that are hard to culture in the vitro, or cannot be cultured at all. Poor quality amplification instruments maintenance and the difference of performance between the enzymes act as a common source of error in the real-time PCR procedures. There is a strong urge to analyze each and every constituent of the experiment on a frequent basis in order to yield results as accurate as possible. Simple errors can be omitted or neutralized by taking certain corrective measures during the study. This will not only strengthen the efficiency of the entire procedures, but will also maximize the probability of accuracy in terms of results and findings. qPCR one of those contemporary biomedical techniques that take place in vitro to allow the exponential amplification of target DNA (Keyu He, 2007).

The multiple regression model considers Ct as the true dependent, β as the intercept, β_xs as the regression coefficients for corresponding X (independent) terms, and ε is the term that represents error. This is represented in the following equation;

$$Ct - \beta 0 + \beta con X icon + \beta treat X itreat + \beta gene X igene + \beta contreat X icon X itrea + \beta congene X icon X igene + \beta genetreat X igene X itreat + \beta congenetreat X icon X itreat X igene + \varepsilon$$

C- Analysis Plan:

The analysis plan consisted of various models and approaches for the reliability of real-time PCR data analysis. The multiple regression model takes into loop the effects of treatment, concentration, gene and their interactions. Here, emphasis has been laid on the interactions between gene and treatment.

An SAS program for multiple regression model "PROC GLM" was applied for the estimation of $\Delta\Delta Ct$ in program 2_MR.sas. in the SAS program, a joint effect of treatment and gene have been evaluated in the estimate and contrast statement, while the multiple regression model has been represented in a model statement. Even though the analysis yielded from the SAS represents quite a comprehensive evaluation of data, only two facets of the analysis were considered.

ANCOVA or Analysis of Covariance

ANCOVA provides researchers with yet another methodology to reach out to the real-time PCR data analysis. This helps in developing a simplified model through converting the data into a grouped data as depicted in the Table 1. The whole procedure results in the following equation;

$$Ct = \beta 0 + \beta con Xicon + \beta group Xigroup + \beta groupcon Xigroup Xicon + \varepsilon$$

Here, the points of interest include the following;

- Adjusted averages of covariance
- Ct difference of target gene value between control sample and treatment

The SAS procedure implemented for the analysis of covariance is pretty much identical to that of the multiple regression model. The procedures applied here could either be PROC GLM or PROC MIXED; while PROC MIXED has been used in this study. The grouping of variables has been identified by the class statement for the purpose of significance testing. The variables considered here are;

- Concentration
- Group

Both of the aforementioned are covariant in nature as assumed by the analysis of covariance.

Even simplified alternatives can be applied for real-time PCR data analysis such as the t-test and the Wilcoxon two group test. Both of these tests are based on the assumption that additional effect of gene, concentration and replicate are adjustable by deducting the Ct number of the target gene from the reference gene. The ΔCt for treatment and control can therefore be subject to simple t-test, which will yield the estimation of $\Delta\Delta$Ct

D- Discussion:

This study presents a more practical approach towards the reliability and data analysis of real-time PCR data, and reflects on its significance and role specifically in biomedical sciences. Comparing the four approaches elaborated above shows some very interesting results and findings. In the case of ANCOVA "Analysis of Covariance" and multiple regression model, both the approaches represent identical results on the grounds of the fact that a similar mathematical approach is implemented for parameter estimation. The T-test and Wilcoxon yield almost similar results with a little difference in the standard error – which is slightly greater for the t-test – while the Wilcoxon two group test yields a relatively smaller estimation of $\Delta\Delta$Ct.

Data Quality Control is a factor of much concern for the real-time PCR data. Simply stating, it has everything to do with estimating the amplification efficiency . In today's world, not many PCR experiments implement a method for measuring amplification efficiency, and also lack a standard curve design. It is proposed that the PCR data analysis conducted in absence of estimation of amplification efficiency is not as reliable as the one conducted in its due presence. Hence, two standards are being put forward for real-time PCR data quality control with the help of SAS programs discussed in the study. These standards are as follows;

1. It is important to include experiments with a serial dilution of DNA templates. This will enable researchers to estimate the amplification efficiency of each gene parallel to the sample
2. It is important to consider experiments with a standard curve

Moreover, it is also essential to consider a real-time PCR instrument system that comprises of three key components including a thermal system, an optical system and software to control the instrument (Suzanne Kennedy, Nick Oswald, 2011).

Conclusion

In this paper, four models were presented for the purpose of statistical analysis of real-time PCR data and a procedure for data quality control. A sample set of data was analyzed through specifically developed SAS programs. It is mandatory to take in to consideration the prevalence of data quality control, and a standard curve in order to ensure highly reliable and valid results of the real-time PCR data analysis. However, despite the fact that real-time PCR is widely under practice by the clinicians for strong positive samples, what happens with the weak positive samples still remains a mystery (Ian Maxwell Mackay, 2007).

References:

E. Van Pelt and Verkuil, "Ensuring PCR Quality – Quality Criteria and Quality Assurance" Springer Science + Business Media B.V, 2008

M Tevfik Dorak, "Real-time PCR" Taylor and Francis Group, 2006

Martin Filion, "Quantitative Real-time PCR in Applied Microbiology", Caister Academic Press UK, 2012

M. Wink, Heidelberg, "Importance of the Polymerase Chain Reaction (PCR) for Plant Sciences", Bioforum Extra, Prague 94, no 5, n.d

Julie Logan, Kirstin Edwards, Nick A. Saunders, "Real-time PCR: Current Technology and Applications", Caister Academic Press UK, 2009

Phouthone Keohavong, Stephen G. Grant, "Molecular Toxicology Protocols" Humana Press Inc. 2005

Suzanne Kennedy, Nick Oswald, "PCR Troubleshooting and Optimization: The Essential Guide" Caister Academic Press UK, 2011

Ian Maxwell Mackay, "Real-time PCR in Microbiology: From Diagnosis to Characterization", Caister Academic Press UK, 2007

Joe O'Connell, "RT-PCR Protocols", Humana Press Inc. 2002

Keyu He, "The Database Implementation and Algorithm Design of QPCR-DAMS: A Database", ProQuest Information and Learning Company, 2007